Practice Papers for SQA Exams

Standard Grade | General

Mathematics

ISBN 978-1-84372-772-9

Published by
Leckie & Leckie Ltd, 4 Queen Street, Edinburgh, EH2 1JE
Tel: 0131 220 6831 Fax: 0131 225 9987
enquiries@leckieandleckie.co.uk   www.leckieandleckie.co.uk

A CIP Catalogue record for this book is available from the British Library.

Leckie & Leckie Ltd is a division of Huveaux plc.

Questions and answers in this book do not emanate from SQA. All of our entirely new and original Practice Papers have been written by experienced authors working directly for the publisher.

Introduction

Layout of the Book

This book contains practice exam papers, which mirror the actual SQA exam as much as possible. The layout, paper colour and question level are all similar to the actual exam that you will sit, so that you are familiar with what the exam paper will look like.

The answer section is at the back of the book. Each answer contains a worked out answer or solution so that you can see how the right answer has been arrived at. The answers also include practical tips on how to tackle certain types of questions, details of how marks are awarded and advice on just what the examiners will be looking for.

Revision advice is provided in this introductory section of the book, so please read on!

How To Use This Book

The Practice Papers can be used in two important ways:

1. You can complete an entire practice paper as preparation for the final exam, sticking to the time allocated and not looking at any notes to help. Mark what you can do in the correct time and see how many marks you got. Next try anything you missed out and see how many more marks you could get if you could work quicker. Then try any remaining questions with the help of your notes and see how many more marks you could have got if you knew the work better!

    This will help you decide how to plan your study.

2. You can use the Topic Index at the front of this book to find all the questions within the book that deal with a specific topic. This allows you to focus specifically on areas that you particularly want to revise or, if you are mid-way through your course, it lets you practise answering exam-style questions for just those topics that you have studied.

Revision Advice

Revising maths means 90% doing questions and 10% looking at your Notes! No one gets good at maths just by reading! If you are also sitting Credit you should use this book to ensure you are at Grade 3 standard (at least 70% of the marks) and then revise Credit for all or most of the time.

Write any important information you need to remember on cards you can carry around with you or store them on your mobile, then you have something useful to look at when you're waiting for a friend to arrive or sitting on a bus.

Maths is very often the first or second exam. If you are making up a revision timetable it needs to have lots of maths on it. Once your exams actually start you will not have to put it in any longer and can concentrate on other subjects.

Work out a revision timetable for each week's work in advance – remember to cover all of your subjects and to leave time for homework and breaks. For example:

Day	6pm–6.45pm	7pm–8pm	8.15pm–9pm	9.15pm–10pm
Monday	Homework	Homework	English revision	Chemistry Revision
Tuesday	Maths Revision	Physics revision	Homework	Free
Wednesday	Geography Revision	Modern Studies Revision	English Revision	French Revision
Thursday	Homework	Maths Revision	Chemistry Revision	Free
Friday	Geography Revision	French Revision	Free	Free
Saturday	Free	Free	Free	Free
Sunday	Modern Studies Revision	Maths Revision	Modern Studies	Homework

Make sure that you have at least one evening free a week to relax, socialise and re-charge your batteries. It also gives your brain a chance to process the information that you have been feeding it all week.

Arrange your study time into one hour or 30 minutes sessions, with a break between sessions e.g. 6 pm–7 pm, 7.15 pm–7.45 pm, 8 pm–9 pm. Try to start studying as early as possible in the evening when your brain is still alert and be aware that the longer you put off starting, the harder it will be to start!

Study a different subject in each session, except for the day before an exam.

Do something different during your breaks between study sessions – have a cup of tea, or listen to some music. Don't let your 15 minutes expand into 20 or 25 minutes though!

Have your class notes and any textbooks available for your revision to hand as well as plenty of blank paper, a pen, etc. You may like to make keyword sheets like the geography example below:

Keyword	Meaning
Anticyclone	An area of high pressure
Secondary Industry	Industries which manufacture things
Erosion	The process of wearing down the landscape

Finally forget or ignore all or some of the advice in this section if you are happy with your present way of studying. Everyone revises differently, so find a way that works for you!

In the Exam

Bring pens/pencils, eraser, ruler, protractor and calculator.

If you have practised a few papers timing yourself you will know whether you normally finish the paper in time or have to rush. If you think you might run out of time, don't rush and don't panic. Work steadily. However, don't linger too long on a question you get stuck on – leave it and go on to the ones you can do. Come back to unfinished questions at the end if you have time.

Never score anything out until you are certain you have put something better in its place! Even if partly wrong, you may get marks for some correct bits. If you do the question again then you can score out the version you think is wrong, but only with one line – don't scribble all over it!

Don't forget there are blank pages at the back if you run out of space. Try to remember to put the number of the question you are doing and it's very helpful to the exam marker if you write next to the original question something like "done at the end".

If you have time at the end check you have done everything and especially that you didn't miss out any part (b) in a question.

Good luck!

Topic	A paper 1	A paper 2	B paper 1	B paper 2	C paper 1	C paper 2
Number calculations, rounding, scientific notation, integers	1, 4	7	1		1, 3, 6	1
Money calculations – bills, foreign exchange, percentage increases, profit	2, 3	1, 5	2	4, 9		3, 7
Time, distance, speed, calendar	2, 8			1	5	1
Algebra	4, 5		7	3	3, 7	
Fractions, percentages, ratio, proportion, variation	6	4	4	4, 5, 7, 10	5	1
Scale, ratio, bearings, similar shapes		8			2, 8	8
Shape, angles, symmetry	9	2	8			2, 4, 6
Geometry of the circle	9	9		6		2, 9
Triangles, trigonometry, Pythagoras		2, 3		8		2, 5
Area and volume		7		6		1, 2
Formulas, rules, patterns, gradient and graph of straight line	7		3, 5, 6	2	4	4
Statistics and graphs, probability	6	6, 10	9	5	9	9, 10, 11

Exam A – General Level

Mathematics

Standard Grade: General

Practice Papers
for SQA Exams

Exam A
General Level
Paper 1
Non-calculator

Fill in these boxes:

Name of centre

Town

Forename(s)

Surname

You are allowed 35 minutes to complete this paper.

You **must not** use a calculator.

Try to answer all of the questions in the time allowed.

Write your answers in the spaces provided, including all of your working.

Full marks will only be awarded where your answer includes any relevant working.

Scotland's leading educational publishers

1. Carry out the following calculations.

(*a*) 452·6 − 238·14

(*b*) $\dfrac{2}{5}$ of 135 grams

(*c*) 4580 ÷ 200

(*d*) 32·7 × 6

2. The cafeteria in Stephanie's school is cashless. On Monday 19th October she starts out with £20 on her card.

She has lunch from the cafeteria on Mondays, Wednesdays and Thursdays each week. Her sandwich and piece of fruit each day comes to a total of £2·30.

KU	RE
1	
2	
1	
1	

			October 2009			
S	**M**	**T**	**W**	**T**	**F**	**S**
				1	2	3
4	5	6	7	8	9	10
11	12	13	14	15	16	17
18	19	20	21	22	23	24
25	26	27	28	29	30	31

Give the **day** and **date** when Stephanie will have to load more money onto her card to be able to buy her usual lunch.

4

3. Massimo's family is visiting his relatives in Italy.

The exchange rate is £1 = 0·97 Euros.

How many Euros does Massimo receive when he exchanges his £80 spending money into Euros?

2

4.

A multi-storey office block has several underground floors for car parking. The lift buttons for the car park floors have negative numbers, so that −1 is one level below the street. Street level is level 0 and the offices begin one floor above at level 1.

KU	RE

Tina parks her car on floor −4 and then rides up 27 levels in the lift to her office. On which floor is her office?

2

5. (*a*) Factorise fully
6*a* + 18*ab*

2

	KU	RE

(b) Solve algebraically
$$8p - 27 = 5p$$

KU: 2

Explorer

6. 20 ~~Venture~~ Scouts are off for a hike and sausage sizzle.

Six of them are vegetarians. 42 meat sausages have been bought for the non-veggie Scouts.

(a) How many vegetarian sausages should be bought so that the vegetarians can each eat exactly the same number of sausages as the others?

KU: 2

(b) The scout in charge of the frying pan has not been paying attention! He put all the meat sausages in but accidentally put one packet of 6 vegetarian sausages in as well.

What is the probability of a scout who takes the first sausage out of the frying pan getting a vegetarian sausage?

Simplify your answer if possible.

RE: 2

7. The diagram shows some designs with square tiles surrounded by rectangular tiles.

(a) Add some black and white tiles to the diagram below to show the design which has 8 black tiles.

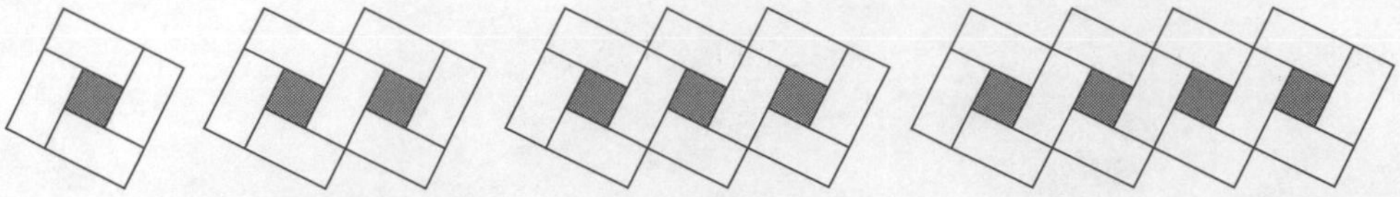

(b) Complete the table below.

Black tiles (b)	1	2	3	4	5	6		12
White tiles (w)	4	7	10					

(c) Write down a formula for calculating the number of white tiles (w), when you know the number of black tiles (b).

(d) Use your formula to calculate the number of black tiles there would be if there were 49 white tiles.

KU | RE

2

2

2

2

8. A flight from Glasgow to Mumbai (India) takes 9 hours 40 minutes.

Local Indian time is 6 hours ahead of Britain.

Rani's family take off from Glasgow at 1935.

What time will it be in Mumbai when the plane lands?

3

9.

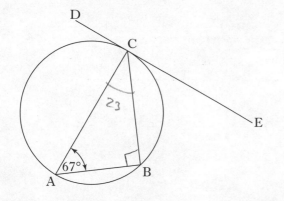

The diagram shows a circle with diameter AC. A tangent to the circle at point C has been drawn. Angle BAC is 67°

1

(*a*) Name one right angle in the diagram

(*b*) Calculate the size of angle ECB.

2

[End of Question Paper]

Mathematics   Standard Grade: General

Practice Papers
for SQA Exams

Exam A
General Level
Paper 2

Fill in these boxes:

Name of centre

Town

Forename(s)

Surname

You are allowed 55 minutes to complete this paper.

You **can** use a calculator.

Try to answer all of the questions in the time allowed.

Write your answers in the spaces provided, including all of your working.

Full marks will only be awarded where your answer includes any relevant working.

<table>
<tr><td></td><td>KU</td><td>RE</td></tr>
</table>

1. Last year Stephanie had £3800 in the bank and she received 6·4% interest on her money at the end of the year.

(a) How much interest did she receive?

2

She withdrew the interest to put towards a holiday and left the original £3800 in the bank for another year.

This year she only receives 1·3% interest.

(b) How much less money will she receive in interest this year?

3

2. AC is a line of symmetry of the rhombus ABCD whose other diagonal BD is 8 units in length.

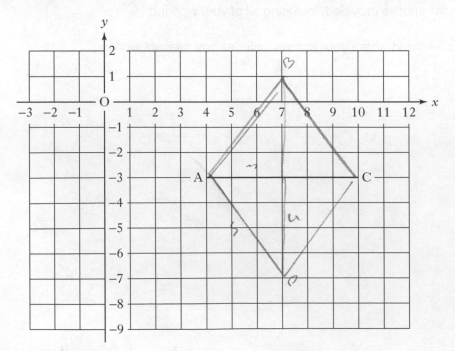

(a) Complete the rhombus on the diagram.

(b) Work out the length of the sides of the rhombus.

3. Angus has a ramp which he attaches to the back of his van to load and unload building materials. He wonders if his ramp is too steep for safety and works out the angle of slope.

To be safe, $x°$ has to be less than 25°. Work out the angle of slope for his ramp. Is Angus's ramp safe?

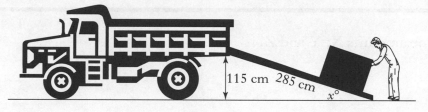

115 cm   285 cm $x°$

4. Daniel wants to buy a new container of Powerclean. He sees that Powerclean comes in two sizes. The assistant says they are equally good value for money.

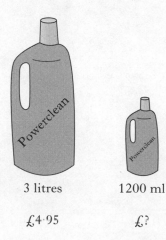

3 litres   1200 ml

£4·95   £?

What is the price of the smaller container?

KU	RE
	2
4	
4	
	3

5. Halima has received her gas bill for the last three months. Here are the charges:

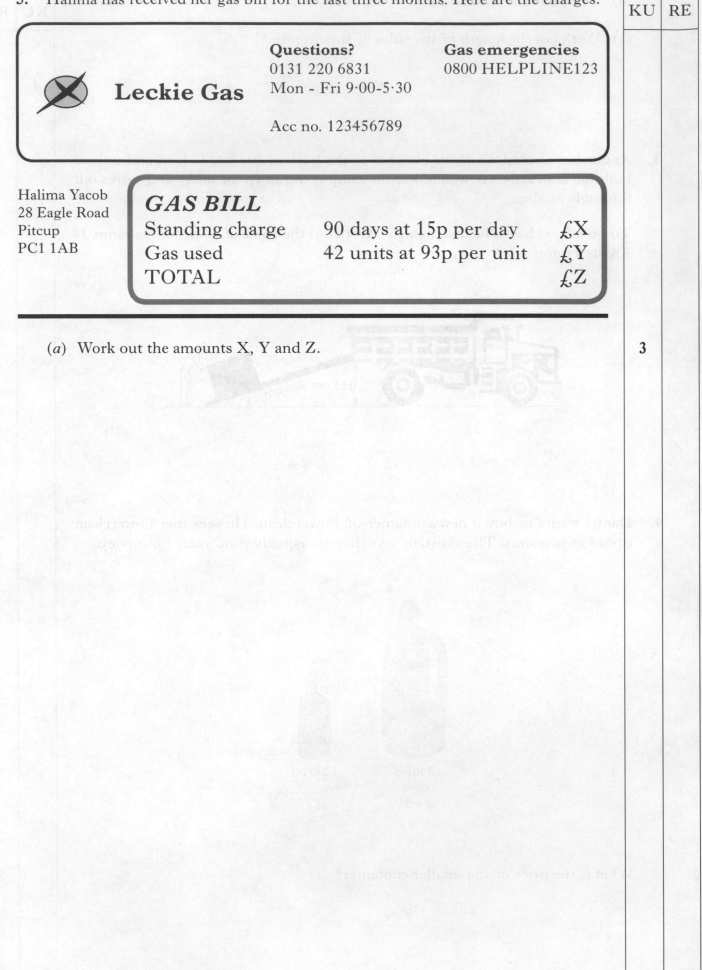

Questions?
0131 220 6831
Mon - Fri 9·00-5·30

Gas emergencies
0800 HELPLINE123

Leckie Gas

Acc no. 123456789

Halima Yacob
28 Eagle Road
Pitcup
PC1 1AB

GAS BILL
Standing charge   90 days at 15p per day   £X
Gas used   42 units at 93p per unit   £Y
TOTAL   £Z

(a) Work out the amounts X, Y and Z.

KU	RE
	3

The graph shows the cost of gas for 90 days on Halima's payment plan (A) and for another supplier where there is no standing charge but each unit costs more (B)

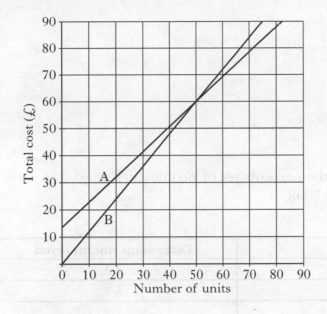

(b) For what number of units used do both plans cost the same?

1

(c) What does each unit cost on plan B?

2

(d) Will Halima save money by switching to this plan, if her usage stays the same? Give a brief reason.

	KU	RE
		1

6. The tables below give the percentages of people unemployed in some European countries in December 2008.

Country	Percentage unemployed
Germany	7·2
Hungary	8·5
Portugal	7·9
Spain	14·4
The Netherlands	2·7
United Kingdom	6·3

The 6 countries in the table were entering **economic recession** towards the end of 2008.

(a) Work out the median and range of percentage unemployment for the countries in the table.

2

(b) Another group of countries were not in recession. Their figures give a median of 6·5% and a range of 6·2%.

	KU	RE

From the information in parts (*a*) and (*b*) of this question do you agree with the statement ......

"If a country is in recession it will have higher unemployment"

Give a short reason for your answer.

7. "Papercom" makes pads of sticky peel-off notes. Each peel-off note is a square of side 8 cm and the height of the block of notes is 8 cm also.

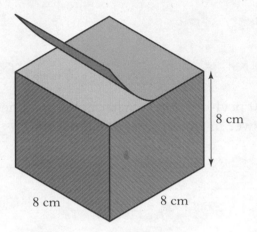

The blocks of notes are packed in cardboard cartons for delivery to the shops. The cartons have internal dimensions 83 cm by 41 cm by 66 cm, as shown in the diagram.

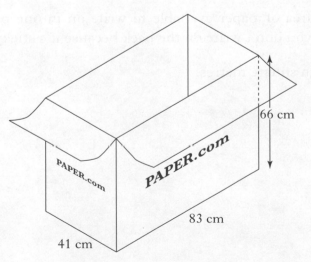

KU	RE
	1

	KU	RE

(a) How many sticky note-pads can be packed in one carton?

RE: 3

(b) Each note-pad has 1000 peel-off notes. What is the thickness of each peel-off note? Give your answer in cm, in scientific notation.

KU: 3

(c) What is the total area of paper available to write on in one pad of sticky notes? Remember you don't write on the back because it's sticky.

Give your answer in square metres.

KU: 3

	KU	RE

8. The diagram shows the positions of Ardrossan and Brodick.

    A ship in the Firth of Clyde is on a bearing of 215° from Ardrossan and 126° from Brodick.

    N

    Ardrossan

    N

    Arran

    Brodick

    Show the position of the ship on the diagram.

    3

9. A restaurant has circular tables of 74 cm high.

    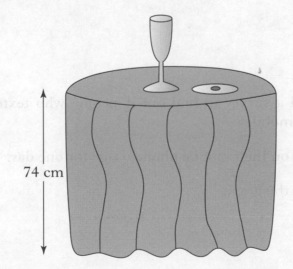

    74 cm

    They have tablecloths which are circular and just touch the floor all the way round.

(a) The diameter of each tablecloth is 290 cm

What is the diameter of each table?

(b) Each tablecloth is trimmed round the edge with red silk. The silk trim costs £3 for each metre.

The bill for the silk trim is £328.

How many tables are there in the restaurant?

10. Two 4th Year pupils did a survey to find out if people who texted most also spoke for longer on their mobiles.

They surveyed 33 people on their mobile phone usage for one day.

The results are shown on the graph.

KU	RE
	2
	4

(a) Draw a best-fitting line on the graph.

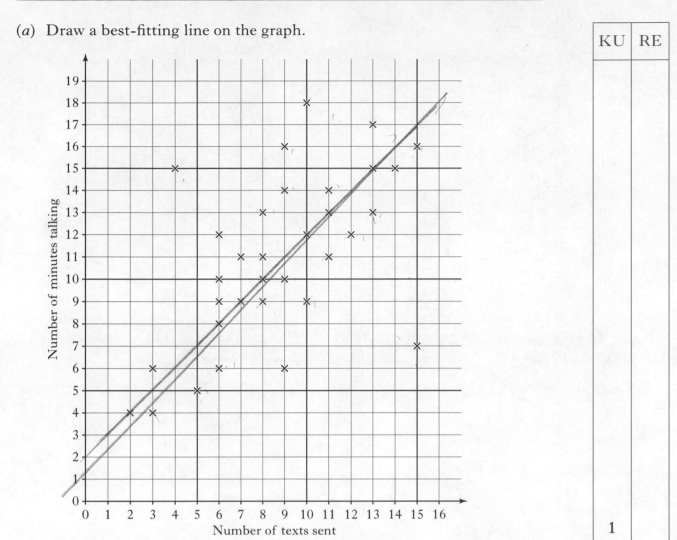

(b) Use your best-fitting line to estimate how many texts would be sent by someone who spoke for 17 minutes on her mobile.

[End of Question Paper]

Exam B – General Level

Mathematics
Standard Grade: General

Practice Papers
for SQA Exams

**Exam B
General Level
Paper 1
Non-calculator**

Fill in these boxes:

Name of centre

Town

Forename(s)

Surname

You are allowed 35 minutes to complete this paper.

You **must not** use a calculator.

Try to answer all of the questions in the time allowed.

Write your answers in the spaces provided, including all of your working.

Full marks will only be awarded where your answer includes any relevant working.

Leckie × Leckie
Scotland's leading educational publishers

	KU	RE

1. Carry out the following calculations.

(a) $34\cdot5 - 18\cdot9 + 55\cdot62$

KU **1**

(b) $17\cdot5 \times 90$

KU **1**

(c) $\dfrac{3}{20}$ of 800

KU **2**

(d) $3 + 8 \times 5$

KU **1**

2. Mr Bruce buys in washing machines at £280 each.

What price should he sell each for to make a profit of 35% on the buying price?

KU **2**

3. (a) Complete all the entries in the table using the rule $y = 2x - 3$

RE **2**

x	0	1	2	3	4	5
y		−1	1			

(b) Plot the points on the graph below and draw a line through them to show the graph of $y = 2x - 3$

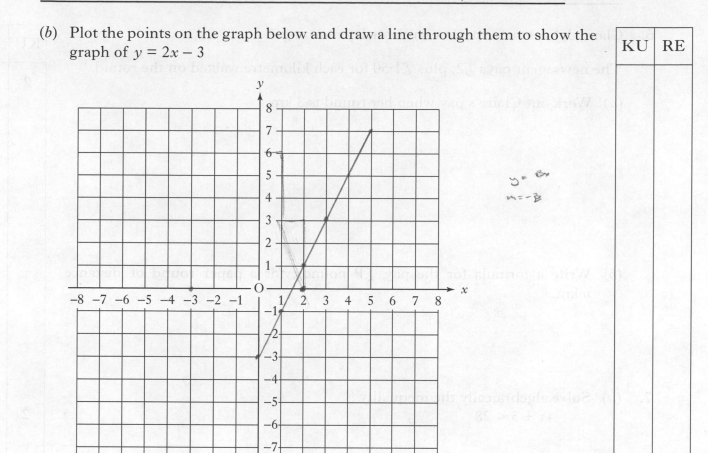

4. Express $37\frac{1}{2}$ % as a fraction in simplest form.

5. The PIN for Declan's mobile phone is a multiple of 5. The PIN has four digits in it. Each of the digits in the PIN number is a different prime number.

Write down all the possible numbers that could be Declan's PIN number.

KU	RE
	3
3	
	3

6. Claire delivers newspapers on Saturday mornings.

The newsagent pays £2, plus £1·50 for each kilometre walked on the round.

(*a*) Work out Claire's pay when her round is 3 km.

(*b*) Write a formula for the pay, £P pounds, for a paper round of distance *n* km.

7. (*a*) Solve algebraically the inequality
$$4x + 5 < 28$$

(*b*) Simplify
$$4a + 6(9a - 2) + 11$$

KU 2

RE 2

KU 2

2

8. Complete this shape so that it has quarter-turn symmetry about O

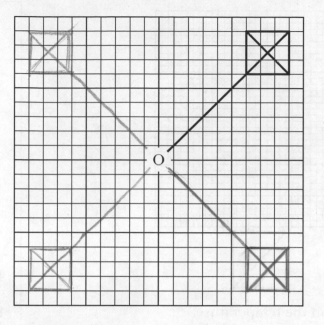

9. The number of hours of sunshine in a Mediterranean holiday resort was logged on 6 successive Wednesdays in the Autumn.

The results are shown in the table below.

Date	16 Sept	23 Sept	30 Sept	7 Oct	14 Oct	21 Oct
Hours of sunshine	8·4	7·6	7·1	6·7	6·9	6·2

(a) Illustrate this data on the grid below using a line graph.

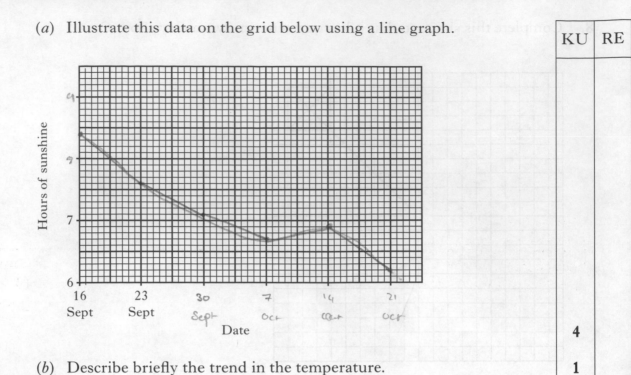

(b) Describe briefly the trend in the temperature.

[End of Question Paper]

Mathematics   Standard Grade: General

Practice Papers
for SQA Exams

Exam B
General Level
Paper 2

Fill in these boxes:

Name of centre

Town

Forename(s)

Surname

You are allowed 55 minutes to complete this paper.

You **can** use a calculator.

Try to answer all of the questions in the time allowed.

Write your answers in the spaces provided, including all of your working.

Full marks will only be awarded where your answer includes any relevant working.

	KU	RE

1. Margaret sets off to drive to Dundee, a distance of 95 km from her home, at 9·24 am. She hopes to average 75 km/h on the journey.

   What time will she arrive if her journey goes according to plan?

   KU: 4

2. A budget shelving system starts with the lowest shelf placed on the floor.

   4 rods slot in at the corners and the next shelf sits on top of these.

   More shelves can be added in the same way, up to the ceiling if desired.

   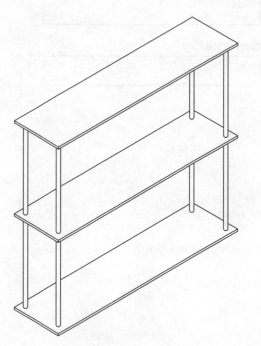

   (a) Complete the table showing how many rods are needed for different numbers of shelves.

   RE: 2

Number of shelves (s)	2	3	4	5	6		10
Number of rods (r)							

	KU	RE

(b) Write down a formula for calculating the number of rods (r), when you know the number of shelves (s)

RE **2**

3. (a) Simplify $7(2x + 5) - 4(x + 2)$

KU **3**

(b) Solve $7n + 5 < 33$

KU **2**

4. Ryan's car has a petrol gauge with 8 divisions. When the petrol in the tank drops to the final division showing that the tank is only $\frac{1}{8}$ full, the display starts flashing a warning and Ryan fills up the tank with petrol.

Petrol costs £0·94 per litre.

The tank holds 48 litres.

What will the cost be for Ryan to fill up his car's petrol tank when the light starts flashing?

RE **3**

	KU	RE

5. Students on a Theatre Arts degree course have to choose which speciality they wish to specialise in. There are four areas students can choose – Dance, Drama, Voice Studies, and Production.

The pie chart shows the breakdown of students in the different specialities in 2010.

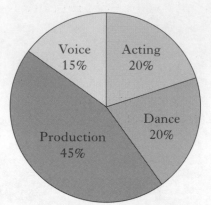

36 students specialise in Voice Studies. How many students are there on the course altogether?

3

6. "Super-Soup" comes in cylindrical cans as shown in the picture. The diagram shows the net of the can.

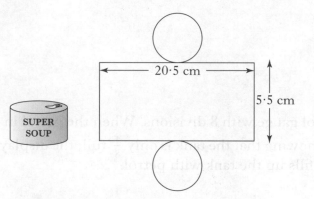

A label covers the curved surface of the can.

(a) What is the area of the label?

2

(b) Work out the area of the lid of the can.

4

7. The current, C amps, of an electrical appliance varies directly with the power rating, P watts, so that

$$P = k \times C \qquad \text{where k is constant}$$

(a) Melissa's hair-straighteners have a power rating of 600 watts. When she plugs them in they deliver a current of 5·5 amps.

Use this information to find the value of k, rounded to the nearest ten.

(b) Use the formula to find what current would be obtained using a foot-spa which has a power rating of 1500 watts.

8. A new design of lampost uses two parallel steel rods to support the horizontal bar with the lamp attached, as shown in the picture.

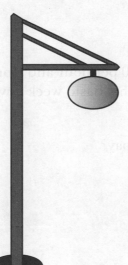

The diagram shows a close-up of the top of the lampost.

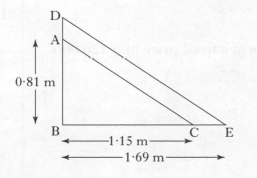

AB = 81 cm
BC = 1·15 metres
BE = 1·69 metres

KU	RE
2	
3	

	KU	RE

(a) AC and DE represent the parallel steel rods.

Calculate the size of angle ACB.

3

(b) The steel for the rods comes in 2 metre lengths.

Will it be possible to cut DE from just one rod?

You must give full reasons for your answer.

5

9. Tom is a joiner. He is paid £16·50 per hour and works a 40 hour week.
To cover holiday pay he is paid his basic week's wages for all 52 weeks of the year.

(a) What is Tom's annual basic pay?

2

Tom's partner, Sarah, is a teacher earning £28 350 annually.

(b) What is the couple's total annual income?

1

(c) The couple want to buy a flat which is for sale at a fixed price of £120 000.

They have savings of £12 600.

The bank will give them a loan calculated using this formula –
Loan amount = 1·75 × total annual income

Can they afford to buy the flat?

You must give full reasons for your answer.

10. Three women are preparing for the opening of their new shop "Jill's Jewels".

They have two large stacks of advertising leaflets to stuff into envelopes and mail out.

They start on the first stack at 1pm and finish at 3 pm.

While they are taking a break, two of their friends turn up and offer to help.

At 3·30 pm they all start on the second stack.

At what time will they finish?

[End of Question Paper]

KU	RE
	2
	4
KU	RE

Exam C – General Level

Mathematics

Standard Grade: General

Practice Papers
for SQA Exams

Exam C
General Level
Paper 1
Non-calculator

Fill in these boxes:

Name of centre

Town

Forename(s)

Surname

You are allowed 35 minutes to complete this paper.

You **must not** use a calculator.

Try to answer all of the questions in the time allowed.

Write your answers in the spaces provided, including all of your working.

Full marks will only be awarded where your answer includes any relevant working.

Scotland's leading educational publishers

	KU	RE

1. Carry out the following calculations.

(a) $72 \cdot 1 + 13 \cdot 99$

KU: 1

(b) 25% of £460

KU: 1

(c) 410×30

KU: 1

(d) $4^2 + 2^3$

KU: 2

2. (a) Write down the length in metres indicated on the scale below.

KU: 1

	KU	RE

(b) To be acceptable the length has to be in the range (1·25 ± 0·001) metres
Is the length in the acceptable range? You must give a reason for your
answer.

RE: 2

3. Eileen went on a very special holiday – a cruise from South America to Antarctica,
where she especially enjoyed seeing icebergs and penguins.

Here is a list of the places she visited, along with some temperatures.

Ushuaia	5°
Falkland Islands	10°
Antarctica	−8°
South Georgia	−2°
Montevideo	.....

(a) How many degrees colder was it in Antartica than in the Falklands?

KU: 2

(b) The temperature measurement at Montevideo was 17° higher than that at
South Georgia. What was the temperature at Montevideo?

KU: 2

4. Ms Davis can never remember her car registration number!

What she does know is that –

It has 2 letters, then 2 numbers, then 3 letters.

The numbers are 54, in that order.

The first 2 letters are F and V

The last 3 letters are G, J and Z

(a) Make a list of all the possible registration numbers it could be supposing that the first two letters are FV in that order.

FV54 ....

FV54 ....

(b) If all the things she remembers are true, how many possibilities will there be altogether?

You must either write down the other possibilities or say why you think your number is correct.

	KU	RE
		3
		2

5. Owen lives next to the Water of Leith walkway where there is a sign to Balerno.

	KU	RE

Balerno $10\frac{3}{4}$ miles   Leith $1\frac{1}{2}$ miles

On Saturday Owen decides to cycle to Balerno and back.

(*a*) How many miles is it to Balerno and back?

1

(*b*) Owen's normal cycling speed is 15 miles/hour on average. How long should it take him to cycle there and back?

3

Owen leaves at 11·30 am. He has 30 minutes of stops for rests and snacks on the way.

(*c*) Will he be back for the beginning of the rugby on TV at 1·30 pm?

2

	KU	RE

6. $3 \cdot 8 \times 10^3 \times A = 3 \cdot 8 \times 10^5$

What is the value of A?

RE: 2

7. (a) Remove the brackets

$$8(3p - 5r)$$

KU: 1

(b) Solve the equation

$$18 - 3x = 12$$

KU: 1

8. Ferry Road is a long straight road – 9 kilometres long. On a map it is drawn as a line measuring 4·5 cm.

(a) Complete this statement giving the scale of the map –

1 cm represents .... km

KU: 1

(b) Work out the scale of the map as a representative fraction, writing it in 1 : .... form.

KU	RE
	3

9. An Animal Rescue service is looking after Lucky, a dog who was very thin when rescued. Lucky's weight is checked regularly.

The results are shown in the table below.

Date	4th Feb	7th Feb	10th Feb	13th Feb	16th Feb	19th Feb
Weight (kg)	28·6	29·4	30·1	30·5	31·2	31·3

Illustrate this data on the grid below using a line graph.

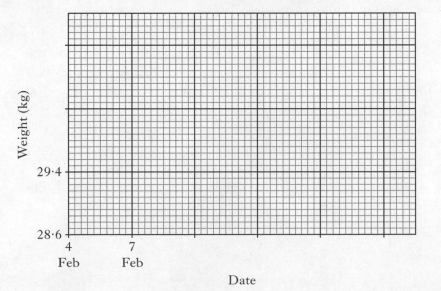

4

[End of Question Paper]

Mathematics   Standard Grade: General

Practice Papers
for SQA Exams

Exam C
General Level
Paper 2

Fill in these boxes:

Name of centre

Town

Forename(s)

Surname

You are allowed 55 minutes to complete this paper.

You **can** use a calculator.

Try to answer all of the questions in the time allowed.

Write your answers in the spaces provided, including all of your working.

Full marks will only be awarded where your answer includes any relevant working.

Leckie×Leckie
Scotland's leading educational publishers

	KU	RE

1. $\dfrac{1}{9}$ of an iceberg lies above the surface of the sea and the rest is below.

The volume of ice above the surface for an iceberg floating in the Scotia Sea is estimated to be 16 000 cubic metres.

(a) Estimate the volume of the iceberg below the surface of the sea?

Write your answer in scientific notation.

3

A ship sighted the iceberg at 2050. Darkness fell, and at 0125 the crew realised they had accidentally come very close to the iceberg and must change direction.

(b) How long was it between first spotting the iceberg and realising they must change direction?

2

2. AB is the diameter of the circle.

C is a point on the circumference of the circle.

AC is 14 centimetres

BC is 5·3 centimetres

(a) Calculate the length of AB, the diameter of the circle.

3

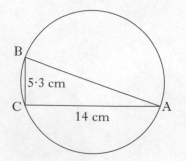

(b) Calculate the area of the triangle ABC.

	KU	RE
	2	

3. Kieran works for a security company. His hourly rate of pay is £13·20 but if he works overtime he is paid time and a half.

(a) Last week he worked 7 hours of overtime.

What will he be paid for the overtime?

2

Kieran takes out a life assurance policy for £150 000.

The annual premium is £1·80 for every £1000 of the life assurance policy.

Kieran pays the premium in monthly instalments.

(b) What is the amount of Kieran's monthly instalment?

2

4. The line AB is drawn on the grid below.

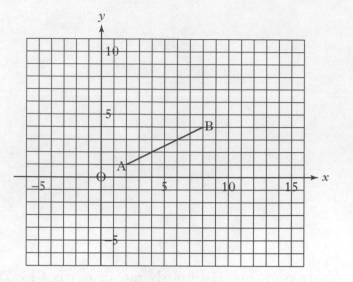

	KU	RE

(a) Calculate the gradient of the line AB.

2

(b) Plot a point C on the diagram such that triangle ABC is isosceles with

AB = BC and the coordinates of C are both integers.

1

(c) Plot a point D on the diagram such that triangle ABD is isosceles with AB = AD and the coordinates of D are both integers.

1

(d) Extend the line AB so that you can see where it cuts the y-axis.

Write down the equation of the straight line which passes through A and B.

2

		KU	RE

5. The clifftop is 150 metres high. The boat is 260 metres away from the foot of the cliff.

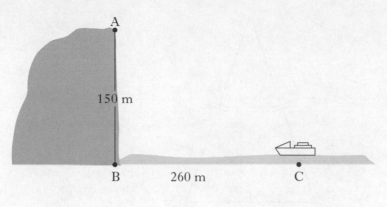

What is the angle of elevation of the cliff top from an observer on the boat?

3

6. (*a*) What is the name of the solid obtained by folding up the net shown in the diagram?

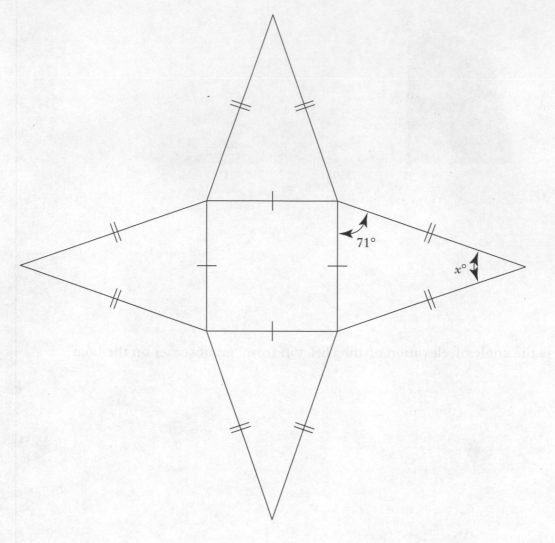

(*b*) Calculate the value of *x* in the diagram.

7. For a cake sale in school some pupils plan to bake a total of 200 cakes. However 12·5% of them are slightly burnt when they come out of the oven.

KU | RE
1 |
| 3

	KU	RE

(a) How many cakes can they sell?

2

The ingredients cost a total of £30·60 and the cakes are priced to cover costs only.

(b) What should they charge for each cake?

2

8. Goldilocks is visiting the Three Bears.

Baby Bear says "Here is my playroom. It's a rectangular shaped room and I have lots of toys all over the floor."

Mummy Bear says "I have a playroom too for my computer and TV and Wii. My playroom is bigger than Baby Bear's but it's similar in shape; the floor is an enlargement of Baby Bear's playroom with a scale factor of 2."

(a) How many times could you fit the floor of Baby Bear's playroom into Mummy Bear's?

2

Daddy Bear says "I have an even bigger playroom and the floor is covered with bits of my car engine. My playroom is an enlargement of Mummy Bear's with scale factor 2."

(b) The area of the floor of Daddy Bear's playroom is 67·84 square metres. What is the area of the floor of Baby Bear's playroom?

3

	KU	RE

9. An old well has a bucket pulled up by a rope which winds round the wheel when the handle is turned.

The diameter of the winding wheel 45 cm.

The bucket must be pulled up 15 metres to be within reach of the person at the top of the well.

How many complete turns of the winding wheel are needed to bring the bucket up?

4

10. SpeedySpin boats run a shuttle speed-boat service to some islands. A boat always waits till at least 14 of the 20 seats on board are occupied.

The table shows the record of passenger numbers for the last 50 journeys.

Number of seats occupied	Frequency	Number of seats occupied x frequency
14	10	
15	12	
16	6	
17	14	
18	5	
19	1	
20	2	
totals	50	

(a) Find the mean number of seats occupied per journey.

Complete the third column to help you.

(b) The operators of the shuttle service claim they are environmentally friendly because they have a 95% occupancy rate. Is this true? Your reason must include a calculation.

11. The stem and leaf chart shows the numbers of pupils in S2 who took up the offer of a free piece of fruit after PE during March.

4	4	7				
5	1	1	3	6	8	9
6	0	1	2			

n = 12

5 | 3 represents 53

All the information in the stem and leaf chart has been displayed correctly except that one data entry has been omitted.

The following statements are true for the <u>correct</u> stem and leaf chart –

Range = 19

median = 54·5

Use this information to decide what the missing entry is.

[End of Question Paper]

WORKED ANSWERS: EXAM A PAPER 1

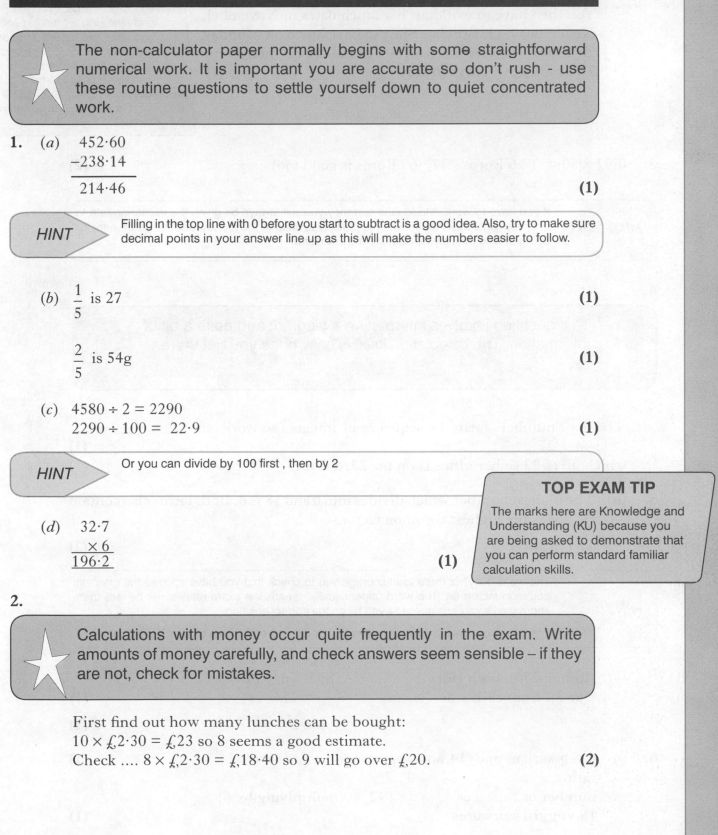

> The non-calculator paper normally begins with some straightforward numerical work. It is important you are accurate so don't rush - use these routine questions to settle yourself down to quiet concentrated work.

1. (a) 452·60
  −238·14
  ———
  214·46 **(1)**

> **HINT** Filling in the top line with 0 before you start to subtract is a good idea. Also, try to make sure decimal points in your answer line up as this will make the numbers easier to follow.

(b) $\frac{1}{5}$ is 27 **(1)**

  $\frac{2}{5}$ is 54g **(1)**

(c) 4580 ÷ 2 = 2290
  2290 ÷ 100 = 22·9 **(1)**

> **HINT** Or you can divide by 100 first , then by 2

TOP EXAM TIP

The marks here are Knowledge and Understanding (KU) because you are being asked to demonstrate that you can perform standard familiar calculation skills.

(d) 32·7
  × 6
  ———
  196·2 **(1)**

2.

> Calculations with money occur quite frequently in the exam. Write amounts of money carefully, and check answers seem sensible – if they are not, check for mistakes.

First find out how many lunches can be bought:
10 × £2·30 = £23 so 8 seems a good estimate.
Check .... 8 × £2·30 = £18·40 so 9 will go over £20. **(2)**

Counting days off on the calendar, there are 6 days she has lunch left in October. **(1)**

You then have to work out her lunch dates in November. November 1st is Sunday. She can have lunch on Monday 2nd and Wednesday 4th, so needs more money on **Thursday 5th November**. **(1)**

> **TOP EXAM TIP**
>
> Reasoning and Enquiry (RE) questions may expect you to think your way through a situation you may not have met before. In question 2 you have to interpret information combining money and dates. To do it successfully you need to be organised with your working and then give the date in full to meet the requirements of the question.

3. $0.97 \times 80 = 77.6$ Euros   (77·60 Euros is good too) **(2)**

> **HINT**
>
> If you have to work with foreign currency you will always be given the exchange rate to use. You get 1 mark for knowing to multiply to change the money, and the other mark for being able to multiply by 80 without a calculator

4.

> ★ This question involves interpreting a diagram and quite a bit of information. The calculation itself is easy once you get there.

The floor numbers form the sequence of integers so work out
$-4 + 27$ **(1)**
which gives 23 so her office is on the **23rd floor**. **(1)**

5. (a) The highest number which divides into 6 and 18 is 6. Both terms also contain "a". 6a is the highest common factor.

   $6a(1 + 3b)$ **(2)**

> **HINT**
>
> The word "fully" is there to encourage you to check that you have spotted the greatest common factor: 6a. The word "algebraically" means the exam marker will be less than impressed if you just guess till you hit on the correct solution.
>
> If you took out only one of the factors, either a 6 or 3 or 2, you would only get 1 mark.

   (b) Adding 27 to both sides   .... $8p = 5p + 27$
       take 5p from each side   .... $3p = 27$ **(1)**
                                            $p = 9$ **(1)**

6. (a) 6 vegetarians and 14 non-vegetarians **(1)**
       ratio   3 : 7
       number of sausages   18 : 42   (multiplying by 6)
       18 veggie sausages **(1)**

(b) There will be 42 meat sausages and 6 veggie in the frying pan.

The chance of picking a veggie sausage is 6 in 48. **(1)**

then simplify   **1 in 8** or $\dfrac{1}{8}$ **(1)**

TOP EXAM TIP

It is important not to miss out second parts of questions just because you are not sure of the first part. Often, as in question 6, you don't need part (a) answer to do part (b).

> **HINT**   Ratio and probability really just involve simple fraction skills.

7.

> Do the diagram in pencil and have a rubber handy – you may discover you need to change something and that's hard if you do it in ink.

(a) If you have the pattern correct but haven't added as many tiles as you should, give yourself one mark. **(2)**

(b)

Square tiles (s)	1	2	3	4	5	6		12
Rectangular tiles (r)	4	7	10	13	16	19		37

(2)

(c)

$$w = 3b + 1$$

> **HINT**   The multiplier is 3 because there is an increase of 3 white tiles for each extra black tile. Then test with the entries in the table to find what to add/subtract to make the formula correct. You would get a mark for "3b" and a mark for "+1"

TOP EXAM TIP

"Use your formula" means you won't receive all the marks for just giving the answer. You are usually being tested on your ability to substitute into a formula – not just writing rows of numbers till you hit on the right one.

(d)

$$w = 3b + 1$$
$$49 = 3b + 1 \qquad \textbf{(1)}$$
$$48 = 3b$$
$$b = 16 \qquad \textbf{(1)}$$

8. 1935 add 9 hours 0435   (be careful counting over midnight)
0435 add 40 minutes 0515
add 6 hours 1115   **(3)**

> **HINT**   There are 3 marks here for writing the arrival time. You don't want to lose all 3 if you make a slight error so write down your working.

9. (*a*) There are three right angles, so you will get your mark for giving any one of them –

ACE, ACD, ABC   **(1)**

> **HINT**   Never be tempted to measure when you are asked to calculate! The drawing will never be drawn to the sizes in the question so you can never get the correct answer that way.

(*b*) 67°

you can get this by finding ACB = 33° from right-angled triangle ABC   **(1)**

and then ACB and BCE make up the right angle ACE, they must add to 90°   **(1)**

> **TOP EXAM TIP**
>
> Make sure there aren't any more questions when you think you have reached the end – the questions are spread over many pages.

WORKED ANSWERS: EXAM A   PAPER 2

1. (*a*) Interest = £3800 × 0·064   = £243·20   **(2)**

(*b*) Interest = £3800 × 0·013   = £49·40   **(2)**

£193·80 less   **(1)**

> **HINT**   You should be competent in changing percentages to decimals by dividing by 100, but you are free to work out percentages such as in this question by any correct method you know.

2. (*a*)

It does not matter which way round B and D are labelled.

                                                          (2)

HINT > Be careful not to rush in and draw a diagram with a pen until you are sure what you are doing! It will be hard to correct the diagram later if you need to. Diagrams are best done in pencil.

(b) $AB^2 = 4^2 + 3^2 = 16 + 9 = 25$ **(3)**

$AB = 5.$ Length of side is 5 cm.   **(1)**

TOP EXAM TIP

Get into the habit of writing in the units in your answer.

HINT > A rhombus is made up of 4 congruent right angled triangles. Use Pythagoras' Theorem

3. $\text{Sin } x° = \dfrac{115}{285} = 0·4035...$ **(1)**

$x = 23·797...$ **(1)**

$x = 23·8$ correct to 1 decimal place   **(1)**

HINT > Make sure your calculator is set for degrees before you even go into the exam! You need $\boxed{\text{inv}}$ or $\boxed{\text{2nd}}$ or $\boxed{\text{sin}^{-1}}$ on your calculator.

TOP EXAM TIP

You would never expect to get 3 marks for writing just "yes it would", would you? Of course not! The point of the question is to get you to show that you know what maths is needed to answer the question and show that you can do the maths.

4. The most efficient calculation is $£4·95 \times \dfrac{1200}{3000} = £1·98$ **(3)**

It is just as good to work out the price for 1ml from the large container and multiply by 1200.

HINT > This is a proportion question but deliberately not made too obvious, which is why you get RE marks for it. You have to realise that the price for each millilitre will be the same if they are equally good value for money.

5. (a) X $90 \times 0·15 = £13·50$ **(1)**
        Y $42 \times 0·93 = £39·06$ **(1)**
        Z   total $£52·56$ **(1)**

(b) 50 units   **(1)**
(This is where the lines intersect)

HINT > In any graph question, be sure you notice what the labels on the graph axes are, and that you read the scale.

(c) cost of 1 unit = $£60 \div 50 = £1·20$ **(2)**

(d) Yes. Plan B is cheaper when fewer than 50 units are bought.   **(1)**

TOP EXAM TIP

1 mark for "brief reason" is to check that you are not just guessing "yes" or "no". Your reason shows that you looked at and understood the graph, and doesn't mean you need to know all about gas bills.

6. (a) median – arranged in order – 2·7 6·3 7·2 7·9 8·5 14·4

7·2 and 7·9 are equally in the middle – find the value midway between them –

$$\frac{7\cdot2+7\cdot9}{2} = 7\cdot55$$ **(2)**

range = 14·4 − 2·7 = 11·7 **(1)**

(b) Yes – the figures show the median unemployment is higher (11·7%) for countries in recession than for the others (6·5%) **(1)**

There are other things you could say instead that would do but this is the simplest – all that is wanted is some evidence you have an idea of what the median and range are and what they tell us.

TOP EXAM TIP

A question worth a mark asking for a comment or an opinion is normally there to check that you have some common sense about what the maths means in real life. For one mark you do not write an essay! – just a short sentence.

HINT You don't need to be an expert on politics or watch News at Ten to answer maths exam questions although it will do you no harm to take a bit of interest. Everything you need to know will be in the question. Notice that part (b) is only for 1 mark.

7. (a) Since 83 ÷ 8 = 10, 41 ÷ 8 = 5 and 66 ÷ 8 = 8 (rounded down) the blocks of notes can be fitted in 10 × 5 × 8 times, that is 400 packs per carton. **(3)**

The really big mistake you could make here is to find the volume of the carton and divide it by the volume of each pad. You have to think!

HINT If you find visualising boxes and cubes tricky do not be afraid to draw sketches you think might help on your script. What you draw can never be counted against you even if it turns out to be irrelevant or wrong.

(b) 8cm depth ÷ 1000 notes **(1)**

= 0·008 cm **(1)**

= 8×10^{-3} cm **(1)**

(c) Each notelet has an area of 8 × 8, or 64 sq cm. **(1)**

There are 1000 in a pad so the total area is 64000 sq cm **(1)**

1 sq m = 100 × 100 sq cm = 10000 sq cm

64 000 sq cm = 6·4 sq m **(1)**

8.

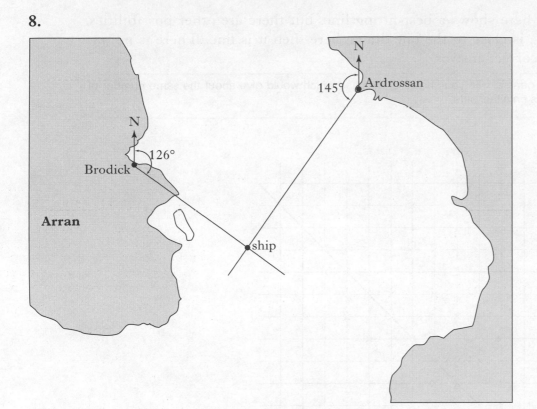

(2)

You would get 1 mark for each of the bearings correct and 1 mark for showing where these lines cross, which is where the ship is.

HINT Don't forget to measure bearings clockwise from North. If you came unstuck on the bearing of 215° then you need to practise a few more drawings using angles over 180° before the exam.

TOP EXAM TIP

Questions involving making an accurate drawing are not asked every year and you can expect one at most. It's best to take your own ruler and protractor to the exam. Remember in these questions accuracy counts – you need to be within 2 degrees for an angle.

9. (a) A line from the floor up to the table top, across the diameter of the table and down again to the floor will measure 74 + d + 74 cm = 290cm **(1)**
So diameter = 290 − 74 − 74 = 142 cm **(1)**

(b) The length of the trim is equal to the circumference of the table cloth.
C = πd
= 3·14 × 2·9 m
= 9·106 m **(1)**
Cost for 1 table = 9·106 × £3 = £27·32 **(1)**

? × £27·32 = £328 .... work out $\frac{328}{27·32}$ = 12
There are 12 tables in the restaurant **(2)**

HINT Visualise what these tables and cloths look like. You don't have to do it in your head - rough sketches will help you see what maths to do.

10. (*a*) The graph here shows a best-fitting line, but there are other possibilities. If your line is close to the one drawn here then it is fine. There is no one absolutely correct answer.

> **HINT** As a general rule you should draw a line which would give about the same number of points on either side.

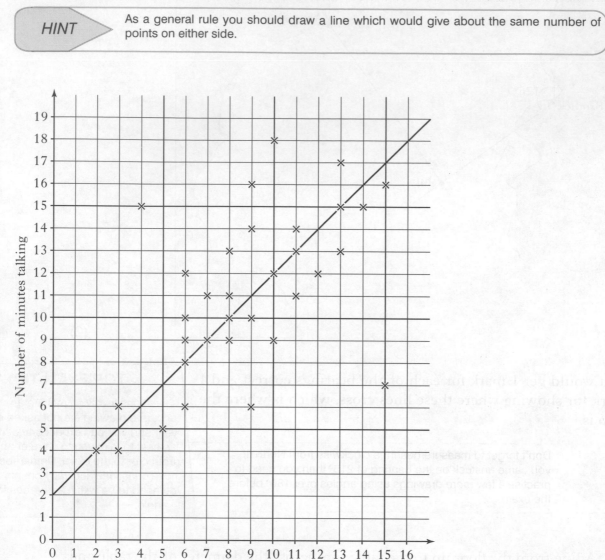

(*b*) Using the line given, the answer is 15 texts.

> **HINT** You must give the answer which fits the line you have drawn.

TOP EXAM TIP

A good use of any spare moments at the end is to check that you have followed the rounding-off instructions in questions that have them.

WORKED ANSWERS: EXAM B   **PAPER 1**

1. Make sure you know your multiplication tables thoroughly – it will make your performance in Paper 1 so much slicker.

    (*a*)

    $$
    \begin{array}{r}
    34{\cdot}5 \\
    -18{\cdot}9 \\
    \hline
    15{\cdot}6
    \end{array}
    \qquad
    \begin{array}{r}
    15{\cdot}6 \\
    +55{\cdot}62 \\
    \hline
    71{\cdot}22
    \end{array}
    $$ **(1)**

    (*b*) $17{\cdot}5 \times 9 = 157{\cdot}5$ **(1)**

          $157{\cdot}5 \times 10 = 1575$

    (*c*) $\dfrac{1}{20}$ is 40 so $\dfrac{3}{20}$ is 120   **(2)**

    (*d*) $3 + 8 \times 5 = 3 + 40 = 43$ **(1)**

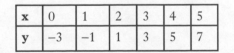

HINT   The multiplication must be done before the addition

TOP EXAM TIP

If you have time at the end some calculations can be easily checked by working backwards.

2. The first way here takes more writing but some people will find it easier to understand

    $10\% = £28$
    $30\% = 3 \times £28 = £84$
    $5\% = £14$
    $35\% = £84 + £14 = £98$
    Selling price $= £280 + £98 = £378$

    You might prefer to calculate $£(280 \div 100 \times 135)$ **(2)**

3. (*a*) 1 mark would be for −3, which is the hardest to do. The other mark would be for all the others.

x	0	1	2	3	4	5
y	−3	−1	1	3	5	7

               (2)

(b)

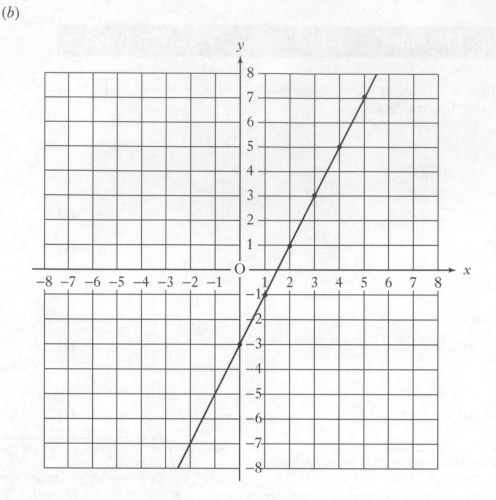

(3)

2 marks for plotting the points (and you would get these even if you had wrong answers in your table, provided where you put them matched your answers) and 1 mark for drawing a straight line neatly through the points. The real line is infinite so it's good to show the line extended a bit at each end. (But you won't get bonus marks!)

HINT

You should draw a neat line with a ruler for the graph in part (b). You should take your own ruler to the exam, but if you forget your centre will make one available for you.

TOP EXAM TIP

For any graph given in the exam paper make sure you have checked the labels are on the axes.

4. Once written as a fraction, eliminate the decimal point first -multiply top and bottom by 10 – before cancelling. Take as many steps as you need and don't stop till you're sure it can't be simplified further.

$$\frac{37 \cdot 5}{100} = \frac{375}{1000} = \frac{75}{200} = \frac{25}{80} = \frac{5}{8}$$

(3)

5. In the exam you are often given a table to write your answers in. Usually you won't be told how many correct answers there are. The table will have enough room for even people who have big handwriting to get them all in. You should carry on till you think you have them all – don't guess how many there are by the space in the table.

2375

2735

3725

3275

7235

7325

(3)

> **HINT**
> Finding the 4 prime numbers 2, 3, 5 and 7 is worth 1 mark. Realising that 5 must be in last place is worth 1 mark. Finding all 6 possibilities is the final mark.

6. (a) Pay = £2 + 3 × £1·50 = £2 + £4·50 = £6·50
   1 mark for 3 × £1·50 and the other for filling in all the rest.

(2)

(b) P = 2 + 1·5n
1 mark for 1·5n. 1 mark for the rest.

(2)

You would expect to get 1 mark for showing that the pay is made up of £2 and £1·5n and another for writing the formula correctly.

> **HINT**
> Part (a) of this question does with numbers what part (b) does with symbols. Make sure you notice what you do with the numbers so you can do the same with the symbols.

> **TOP EXAM TIP**
> Formulas should not have units, for example "km" or "£" written in them.

7. (a) $4x < 28 - 5$ (taking 5 from each side)
             $4x < 23$ 1 mark for getting this far

             $x < 5\dfrac{3}{4}$ or $x < 5{\cdot}75$ 1 mark for either fraction or decimal version

    (2)

    (b) $4a + 54a - 12 + 11$ 1 mark (multiplying out brackets)
             $58a - 1$ 1 mark (collecting terms)

    (2)

> **HINT**   Watch out when collecting terms where some are negative. Also be careful that you have used the inequality sign in the answer.

8.

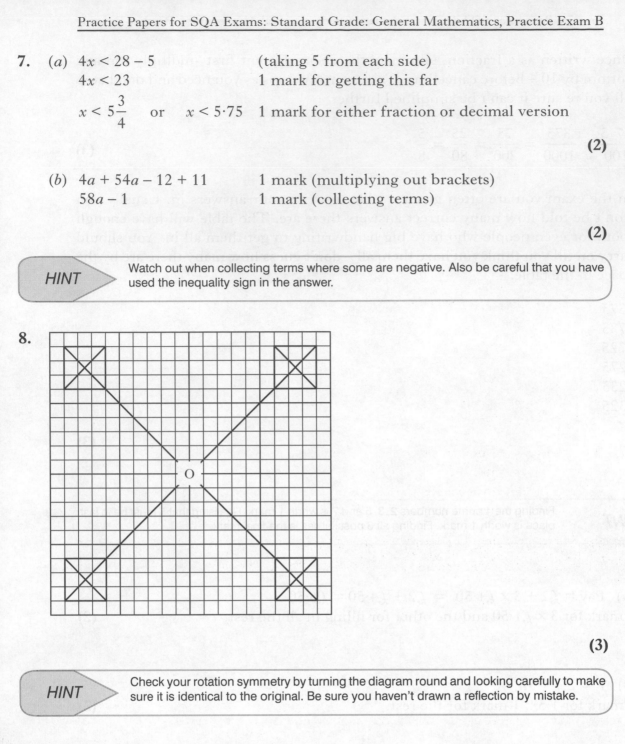

(3)

> **HINT**   Check your rotation symmetry by turning the diagram round and looking carefully to make sure it is identical to the original. Be sure you haven't drawn a reflection by mistake.

9.

> **HINT**   Use the values on the scale given and work out how to spread the other values you need out. Check what the highest values you need are before you start labelling, and avoid the graph being crushed up.

(a)

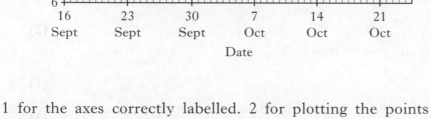

(4)

1 for the axes correctly labelled. 2 for plotting the points correctly.
1 for joining the points

(b) The number of **hours** of sunshine goes down as **time** passes. Alternatively, a passable answer would be 'The trend is downwards', or, 'the temperatures fall.' No need to say more for just 1 mark. **(1)**

WORKED ANSWERS: EXAM B   PAPER 2

1. Time for journey = $\dfrac{95}{75}$ hours = 1·266666 ... h   **(1)**

$$= 1 \text{ hr } 16 \text{ mins}$$ **(2)**

On calculator, deduct 1 (leaving 0·2666) then multiply by 60 to change the decimal into minutes, giving 16 minutes. So the journey takes 1 h 16 mins.
Arrival time = 10·40 am (1 hr 16 mins after departure time)   **(1)**

HINT ▷ The big mistake is to think that, for example, 3·42 hours = 3 hours 42 minutes. Don't fall into that trap.

2. (*a*)

Number of shelves (s)	2	3	4	5	6		10
Number of rods (r)	4	8	12	16	20		36

1 mark for "36", and 1 mark for all the others

(*b*) $r = 4s - 4$ **(2)**

> **HINT** For the rule, look for a "multiply by" and then an "add on" (which might be a "take off" instead)

3. (*a*) $14x + 35 - 4x - 8$ **(2)**
(take care with signs in second bracket)

     $= 10x + 27$ **(1)**

(*b*) $7n < 28$ **(1)**
(by taking 5 from each side)

     $n < 4$ **(1)**

> **HINT** It's important that you use the correct sign ($<$) in your answer for part (*b*).

4. Petrol left in tank $= \dfrac{1}{8}$ of 48 litres $= 6$ litres   **(1)**

petrol needed to top up $= \dfrac{7}{8}$ of 48 litres $= 7 \times 6$ litres $= 42$ litres   **(1)**

cost of petrol $= 42 \times £0{\cdot}94 = £39{\cdot}48$ **(1)**

> **HINT** Each piece of information here is needed, and it is your job to see how they link together. It can help to write the facts down in a list and look for the links between them.

5. $15\% = 36$ students   **(1)**
$5\% = 36 \div 3 = 12$ **(1)**
$100\% = 12 \times 20 = 240$ students   **(1)**

> **HINT** You can solve this question by finding out how many students make 1%, then 100%, but the method here works out 5% instead – easier numbers and just as good, but it's your choice.

6. (*a*) Area $= 20{\cdot}5 \times 5{\cdot}5$ **(1)**
         $= 112{\cdot}75$ sq cm   **(1)**

> **HINT** The rectangle on the net is the part of the can to which the label is glued.

(b)

$$C = \pi \, d$$
$$20 \cdot 5 = \pi \, d \qquad \textbf{(1)}$$
$$d = 20 \cdot 5 \div \pi$$
$$= 20 \cdot 5 \div 3 \cdot 14$$
$$= 6 \cdot 525 \text{ cm} \qquad \textbf{(1)}$$
$$\text{Area} = \pi \, r^2 \qquad \textbf{(1)}$$
$$= \pi \times 3 \cdot 263^2$$
$$= 33 \cdot 4 \text{ cm}^2 \qquad \textbf{(1)}$$

HINT The long side of the rectangle must be the same length as the circumference of the lid.

TOP EXAM TIP

Formulas you might need to use are given at the beginning of the exam paper. Check you get them right.

7.

You don't need to know anything about electricity to do this question. You don't need to have used hair-straighteners either! Good physicists might like to know that Melissa would be in America rather than Britain for this formula to be correct. Think about it!

(a) $600 = k \times 5 \cdot 5$ **(1)**

$$k = \frac{600}{5 \cdot 5} = 110 \text{ (to nearest ten)} \qquad \textbf{(1)}$$

(b) The formula becomes $P = 110C$ **(1)**

substituting $1500 = 110C$ **(1)**

solving $C = \dfrac{1500}{110} = 13 \cdot 6 \text{ amps}$ **(1)**

8.

In a maths exam "reasons for your answer" usually means showing calculations. You can't expect 5 marks without some good maths written down.

(a)

$$\tan A\hat{C}B = \frac{AB}{BC} = \frac{0 \cdot 81}{1 \cdot 15} \qquad \textbf{(2)}$$

$$A\hat{C}B = 35 \cdot 2^\circ \text{ (using inverse tan)} \qquad \textbf{(1)}$$

(b)

HINT There are two ways you might think of to do part (b). This is fine – you have the choice. Part of the skill of mathematics is to explore different methods and make choices

<u>1st method</u>

Angle DEB = angle ACB   (because AC and DE are parallel)   **(1)**

Using triangle BDE, $\cos BED = \dfrac{BE}{DE}$ **(1)**

$$\cos 35 \cdot 2^\circ = \dfrac{1 \cdot 69}{DE}$$ **(1)**

$$DE = \dfrac{1 \cdot 69}{\cos 35 \cdot 2} = 2 \cdot 07 \, m$$ **(1)**

so no, it won't be possible to make DE from 1 rod.

<u>2nd method</u>

find AC using Pythagoras' theorem –   **(1)**
    $AC^2 = AB^2 + BC^2 = 0 \cdot 81^2 + 1 \cdot 15^2 = 1 \cdot 9786$

    $AC = \sqrt{1 \cdot 9786} = 1 \cdot 41 m$ **(1)**

Since triangles ABC and DBE are similar, $\dfrac{DE}{AC} = \dfrac{BE}{BC}$ **(1)**

$$\dfrac{DE}{1 \cdot 41} = \dfrac{1 \cdot 69}{1 \cdot 15}$$

$$DE = 2 \cdot 07 \text{ m}$$ **(1)**

so no, it won't be possible to make DE from 1 rod.   **(1)**

TOP EXAM TIP

Exam markers, when giving RE marks, are trying to look especially for your good ideas on how to solve the problem, and are not looking so hard at whether all your working is correct. Make sure it's clear what methods you are using.

9.

Exam questions are checked carefully to make sure words used in questions about houses, banks, loans etc are likely to be familiar to candidates. Make sure that you pay attention in class when your teacher tells you about these words, even if at your age you still find mortgages and rent rather boring. One day it will probably become very interesting to you.

(*a*)   Annual pay $= 52 \times 40 \times £16 \cdot 50$ **(1)**
          $= £34 \, 320$ **(1)**

(*b*)   total $= £34 \, 320 + £28 \, 350 = £62 \, 670$ **(1)**

(c)   Loan = $1 \cdot 75 \times £62\,670 = £109\,672 \cdot 50$ **(1)**
        $£109\,672 \cdot 50 + £12\,600 = £122\,272 \cdot 50$
and as this is greater than the price of the house they can afford the flat.   **(1)**

> **HINT**   This is a standard method for inverse proportion questions but it is not the only way it can be done.

people	hours		
3	2		
1	$2 \times 3 = 6$	(takes 3 times as long on your own)	**(1)**
5	$6 \div 5 = 1 \cdot 2$	(more people, less time)	**(1)**

$0 \cdot 2$ hours $= 0 \cdot 2 \times 60 = 12$ mins so it takes 1 hour 12 minutes   **(1)**
They will finish at 4·42 pm   **(1)**

> **TOP EXAM TIP**
> Always write down enough working to get some of the marks at least if you go a little bit wrong. It does not matter how you set down working so long as it is comprehensible to the reader.

WORKED ANSWERS: EXAM C   PAPER 1

1. (a)
$$\begin{array}{r} 72{\cdot}1 \\ +\,13{\cdot}99 \\ \hline 86{\cdot}09 \end{array}$$

(b)   divide by 4 ....   £115

(c) $410 \times 3 = 1230$
$1230 \times 10 = 12300$

(d) $4^2 = 4 \times 4 = 16$ $2^3 = 2 \times 2 \times 2 = 8$
answer is 24

> **HINT** — 1 mark would be for showing that you know what a power means – by working at least one of them out correctly. The other mark would be for getting all the rest right too.

> **HINT** — Be careful not to confuse
> $2 \times 2 \times 2 = 8$ with
> $2 + 2 + 2 = 6$

2. (a) $1{\cdot}248$

(b) $1{\cdot}25 + 0{\cdot}001 = 1{\cdot}251$ $1{\cdot}25 - 0{\cdot}001 = 1{\cdot}249$ **(1)**
Since $1{\cdot}248$ isn't between them it is not acceptable.   **(1)**

> **HINT** — If you find working out the numbers in between difficult, write the marked numbers as $1{\cdot}240$, $1{\cdot}250$, etc, and it will be easier.

TOP EXAM TIP

It's very unlikely that "yes" or "no" will be enough even for a 1 mark question. You would have a 50% chance of being correct even guessing. Always explain.

3.

> ★ This question is testing your knowledge of integers (positive and negative numbers). If you find it a help, draw a number line showing positive and negative numbers to help you calculate.

(a)   From 10 down to -8 is 18 degrees

(1 mark is for finding the numbers -8 and 10 from the list. 1 mark for finding the difference between them)

(b)   Start from -2 and go up 17   **(1)**
$-2 + 17 = 15$ **(1)**

TOP EXAM TIP

You will never lose marks for drawing number lines, diagrams.... to help you, even if they are wrong or not relevant at all. You get marks for everything that can be found that's correct, and anything else just doesn't matter. So if you think a diagram MIGHT help.... then go for it!

4.

(*a*)   After "FV54" you can have all of the following -
     GJZ   GZJ   JGZ   JZG   ZGJ   ZJG
     so you should have 6 possibilities.

(1 mark for every 2 correct)

(*b*)   There are 12 possibilities altogether.   **(1)**
     If the V and F are the other way round there will be another 6 possibilities –
     VF54 followed by each of the 6 combinations listed in (*a*). That means 12
     possibilities altogether.   **(1)**

     Alternatively, you could write out the extra 6 instead – VF54 GJZ, VF54
     JGZ and so on (1) and then say that makes 12 altogether. (1)

5.

(*a*) $2 \times 10 = 20$ $2 \times \dfrac{3}{4} = 1\dfrac{1}{2}$ total distance $= 21\dfrac{1}{2}$ miles   **(1)**

(*b*)   1 hour for the first 15 miles. $6\dfrac{1}{2}$ miles left.
     60 minutes for 15 miles means 4 minutes per mile.
     $6 \times 4 = 24$ mins $\dfrac{1}{2} \times 4 = 2$ mins   total time $= 1$ hr 26 mins

(3)

(*c*)   Including stops he will take 1 hour 56 mins so should arrive back with 4
     minutes to spare, so yes, he should make the rugby.   **(2)**

6. $3800 \times A = 380000$ **(1)**
       so $A = 100$ **(1)**
           or 10^2

HINT You can write down the answer very easily with no working at all if you really understand what scientific notation is, but it can be worked out by writing the numbers in full if not.

7. (a) 24p − 40r **(1)**

(b) $18 - 6 = 12$ so $3x = 6$, **(1)**
$x = 2$ **(1)**

HINT This equation is probably easier to solve by looking to see what number must replace "3x" than by any other method. Or, if you prefer, use your usual method.

TOP EXAM TIP

Unless you are told you must use a particular method, the choice is yours. However, the exam is not the best place to try any wacky new methods of your own.

8.

This question reminds you that you need to know the number of centimetres in a metre, metres in a kilometre, millilitres in a litre and so on.

(a) 4·5 cm represents 9 km
1 cm represents 2 km **(1)**

(b) 1 cm represents 2×1000 m = 2000 m **(1)**
1 cm represents 2000×100 cm = 200 000 cm **(1)**
1 : 200 000 **(1)**

9.

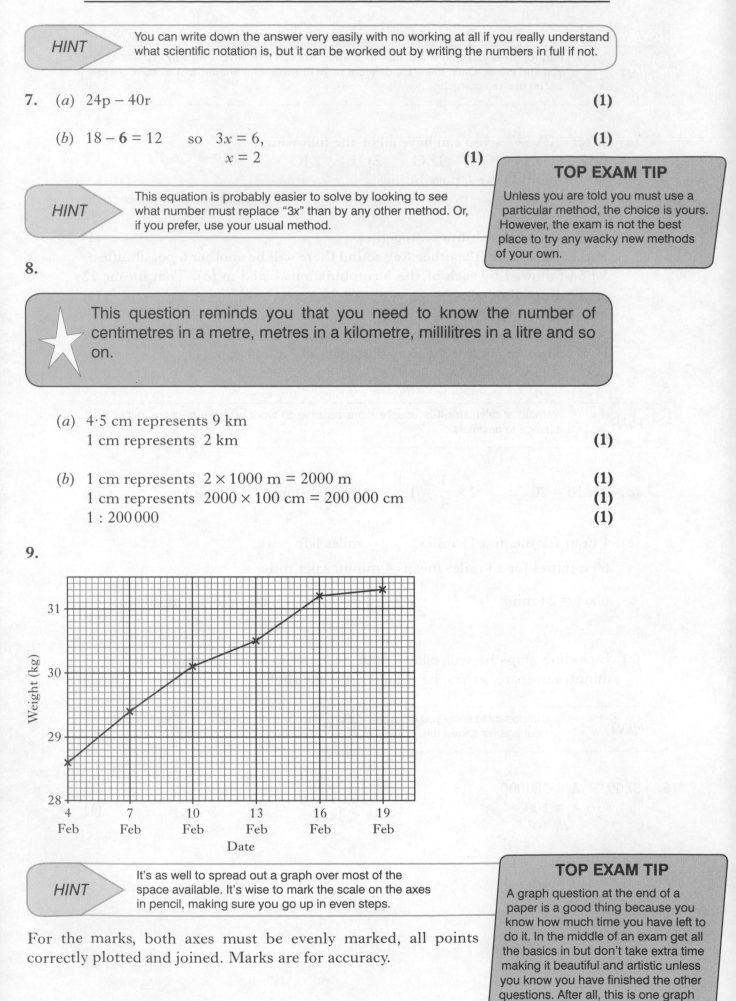

HINT It's as well to spread out a graph over most of the space available. It's wise to mark the scale on the axes in pencil, making sure you go up in even steps.

For the marks, both axes must be evenly marked, all points correctly plotted and joined. Marks are for accuracy.

TOP EXAM TIP

A graph question at the end of a paper is a good thing because you know how much time you have left to do it. In the middle of an exam get all the basics in but don't take extra time making it beautiful and artistic unless you know you have finished the other questions. After all, this is one graph which is **never** going to be pinned up on your teacher's wall!

| WORKED ANSWERS: EXAM C | PAPER 2 |

1. (a) $\frac{1}{9}$ above so $\frac{8}{9}$ below **(1)**

$8 \times 16\,000 = 128\,000$ cubic metres **(1)**
$$= 1{\cdot}28 \times 10^5$$ **(1)**

> **HINT** Notice that the question doesn't ask for the **total** volume of the iceberg.

(b) There are different ways to calculate the time interval. Here's one ...

2050 – midnight is 3 hours 10 minutes.

Add 1 hr 25 minutes to get 4 hours 35 minutes. **(2)**

2. (a) Since AB is a diameter of the circle, angle ACB is 90° **(1)**

$$
\begin{aligned}
AB^2 &= AC^2 + BC^2 \\
&= 14^2 + 5{\cdot}3^2 \\
&= 196 + 28{\cdot}09 \\
&= 224{\cdot}09
\end{aligned}
$$ **(1)**

$$AB = \sqrt{224{\cdot}09} = 14{\cdot}97.... = 15 \text{ cm}$$ **(1)**

(b) Area of right angled triangle $= \frac{1}{2}$ base × height

$$= \frac{1}{2} \times 14 \times 5{\cdot}3$$ **(1)**

$$= 37{\cdot}1 \text{ sq cm}$$ **(1)**

> **HINT** Diameters of circles should make you think of right-angled triangles in the semicircles. This means you can use trigonometry and Pythagoras theorem.

TOP EXAM TIP

Make sure you do every part of a question – most have more than one part and the later parts often have more marks.

3. (*a*) overtime rate = $1.5 \times £13.20$ $= £19.80$ **(1)**
overtime pay = $7 \times £19.80$ $= £138.60$ **(1)**

(*b*) annual premium = $150 \times £1.80 = £270$ **(1)**
monthly payment = $£270 \div 12 = £22.50$ **(1)**

> **HINT** Make sure you are familiar with words like premium, annual, overtime – they may not have much relevance to your life as yet but you have to show you are prepared for adult life.

4.

> ★ This sort of question asks you to experiment a bit. There are several correct positions in parts (*b*) and (*c*) but you do not need to find them all.

(*a*) Gradient = $\dfrac{vertical}{horizontal} = \dfrac{3}{6} = \dfrac{1}{2}$ or 0.5 **(2)**

> **HINT** It doesn't matter which way you give the answer. You would get a mark for getting 3 and 6 for the distances and the other mark for working out the gradient

(*b*) There are six places C could be plotted: (2, 7) and (14, 1) are probably the most obvious but (5, 10), (5, −2), (11, −2) and (11, 10) are also correct. The mark is given for any one of these.

(*c*) Similarly there are six correct answers for this: (− 4, 4), (8, −2), (−1, 7), (5, 7), (5, −5), and (−1, −5). Any one of these gets the mark.

(*d*) The intercept on the y-axis is 0. **(1)**
Substituting into $y = mx + c$ gives

$y = \frac{1}{2}x$, or $y = 0.5x$, (or $2y = x$ if you multiply through to remove the fraction).

Any of these gets the marks.

> **TOP EXAM TIP**
> If you leave a question unfinished, put a mark beside it to remind you to look at it again if you have time at the end.

5. The angle of elevation is $B\hat{C}A$ **(1)**

$\tan B\hat{C}A = \dfrac{150}{260} = 0.5769....$ **(1)**

$B\hat{C}A = 30°$ **(1)**

> **HINT** Suppose you forgot what the angle of elevation meant. Then you could guess which angle, write it down and work out the answer for that angle. You won't get the mark for knowing the correct angle, but if you find your angle correctly you will likely get the two marks for showing you can do trig.

6. (a) square-based pyramid or square pyramid **(1)**

> **HINT** Give the name of the shape in full if you have the option. Extra information can be ignored but something missing loses you marks.

(b) to fold up correctly the triangles are congruent isosceles triangles.
$2 \times 71 + x = 180$ the angle sum in the triangle **(1)**
so $x = 28$ **(1)**

7. (a) $12 \cdot 5\%$ is $\frac{1}{8}$ so divide 200 by 8 ...... **(1)**

25 burnt so 175 for sale. **(1)**

(b) £30·60 ÷ 175 = 0·17485..... **(1)**

charge of 18 pence each **(1)**

> **HINT** Sometimes you have to round sensibly, taking into account the situation in the question, as in the second part here.

8.

> ★ It's a kids' story but this maths isn't kids' stuff!

The question is about similarity and scale factors. Sketches of the rectangular floors would help you see what to do if you weren't sure.

(a) scale factor for lengths = 2 **(1)**
scale factor for area = $2^2 = 4$, so it would fit in 4 times. **(1)**

(b) Similarly, the floor of Daddy Bear's playroom is 4 times as large as Mummy Bear's. **(1)**

That means it is 4×4 as big as Baby Bear's, that is, 16 times as big **(1)**
Area of Baby Bear's floor = Area of Daddy Bear's floor ÷ 16
= 67·84 ÷ 16 = 4·24 sq m **(1)**

> **TOP EXAM TIP**
>
> It is not the case that questions with lots of words are harder – in fact it is often the opposite. Do not be put off. If words are a problem for you then extra time or a reader in the exam will probably have been arranged for you.

9. circumference of wheel = $\pi \times d = 3 \cdot 14 \times 45 = 141 \cdot 3$ cm **(1)**

number of turns needed $= \dfrac{15 metres}{141 \cdot 3cm} = \dfrac{1500 cm}{141 \cdot 3cm} = 10 \cdot 6...$ **(2)**

number of complete turns = 11 **(1)**

> **HINT** One turn of the wheel winds a length of rope equal to the circumference round the wheel, and the bucket rises by that same amount.

> **TOP EXAM TIP**
>
> Question 9 is the sort of question (RE) where more marks will be given for doing the right things, and less for getting every calculation correct.

10. (*a*) The entries in the third column, reading downwards, are – 140, 180, 96, 238, 90, 19, 40 **(1)**
and the total is 803 **(1)**

Mean $= \dfrac{803}{50} = 16 \cdot 06$ **(1)**

(*b*) 95% of 20 seats = 19 **(1)**
As this is higher than 16 it is not true that they have 95% occupancy. **(1)**

 HINT Make sure you are quite clear on the difference between the different averages which can be calculated – mean, median and mode.

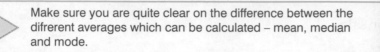

TOP EXAM TIP

Questions on topics like Green issues might make you want to express opinions about the issues in the question. You will not lose marks for adding extra comments of your own, though you won't gain any either! For example, in question 10, although the 95% claim is not true, the occupancy rate is still very high and the shuttle service does seem to be acting in an environmentally responsible way and you might wish to say that.

In many controversial issues Mathematics is used by both sides to justify their claims. That is one reason you learn Mathematics – to learn how to back up claims about important real-life issues with solid facts.

11.

HINT It's easy enough to put all the information in a table (provided you're careful) but this question makes you think a bit more.

The range of the data given = 62 – 44 = 18   but should be 19.
This means the missing entry is either 43 or 63. **(1)**

If 43 is added, the two pieces of data in the middle are 53 and 56

median $= \dfrac{53+56}{2} = 54 \cdot 5$

If 63 is added, the two pieces of data in the middle are 56 and 58, giving a median of 57   (1 mark for being able to calculate a median correctly)

So **63** is the missing piece of data **(1)**